Preface

Welcome to the Science Olympiad All Chapters Combined MCQ Class 10!

This book is specially crafted to ignite the spark of curiosity and foster a love for science in young minds. Designed with the unique needs of first graders in mind, it provides a comprehensive introduction to fundamental scientific concepts through engaging, age-appropriate content.

Our primary goal is to make learning science an enjoyable and enriching experience. The book is filled with colorful illustrations, real-life examples, and interactive exercises that help children understand and relate to the world around them. Each chapter is carefully structured to build on prior knowledge, ensuring a steady progression in learning.

Key features of this book include:

Interactive Exercises: Hands-on activities and simple experiments designed to engage students actively and make learning fun.

Multiple Choice Questions: End-of-chapter MCQs help reinforce learning and provide ample practice for young learners.

Cross-Curricular Connections: Integrates elements of math and language arts, promoting a holistic learning experience.

Environmental Awareness: Introduction to basic environmental science and sustainability concepts.

Through this book, we aim to encourage critical thinking, problem-solving skills, and a deeper understanding of scientific principles. The inclusion of real-life scientists and their contributions will inspire young learners and highlight the importance of science in our daily lives.

Whether used in the classroom or at home, this book is an invaluable resource for fostering a lifelong love of science. We hope that it will be a stepping stone for students, helping them explore and appreciate the wonders of the scientific world.

Thank you for choosing our Science Olympiad Book. Let's embark on this fascinating journey of discovery and learning together!

Happy Learning!

Table of Contents

Questions: Part 1

1. Which of the following is a balanced chemical equation?

a) $H_2 + O_2 \rightarrow H_2O$

b) $2H_2 + O_2 \rightarrow 2H_2O$

c) $H_2 + O_2 \rightarrow 2H_2O_2$

d) $H_2O_2 \rightarrow H_2 + O_2$

2. What is the pH of a neutral solution at 25°C?

a) 0

b) 7

c) 14

d) 10

3. Which of the following metals is the most reactive?

a) Copper

b) Iron

c) Sodium

d) Gold

4. Which allotrope of carbon is used in pencils?

a) Diamond

b) Graphite

c) Fullerene

d) Carbon black

5. Which process in living organisms involves the conversion of glucose into energy?

a) Photosynthesis

b) Respiration

c) Digestion

d) Excretion

6. Which hormone is responsible for regulating blood sugar levels?

a) Insulin

b) Adrenaline

c) Thyroxine

d) Glucagon

7. Which method of reproduction is observed in bacteria?

a) Binary fission

b) Budding

c) Fragmentation

d) Spore formation

8. Who is known as the 'Father of Genetics'?

a) Charles Darwin

b) Gregor Mendel

c) Louis Pasteur

d) Alexander Fleming

9. Which law explains the phenomenon of reflection?

a) Ohm's Law

b) Newton’s First Law

c) Law of Reflection

d) Boyle’s Law

10. What is the main function of the human eye?

a) Hearing

b) Seeing

c) Smelling

d) Tasting

11. Which unit is used to measure electric current?

a) Volt

b) Ampere

c) Ohm

d) Coulomb

12. What is produced when a current-carrying wire is placed in a magnetic field?

a) Electric force

b) Magnetic force

c) Gravitational force

d) Nuclear force

13. Which process helps in the recycling of nutrients in an ecosystem?

a) Photosynthesis

b) Decomposition

c) Respiration

d) Evaporation

14. Which of the following does not change during a chemical reaction?

a) Mass

b) Number of molecules

c) Number of atoms

d) Volume

15. Which is a property of an acid?

a) Bitter taste

b) Slippery feel

c) Sour taste

d) Turns red litmus blue

16. What is the chemical formula of quicklime?

a) CaO

b) $CaCO_3$

c) $Ca(OH)_2$

d) $CaCl_2$

17. Which compound is used as an antifreeze in car radiators?

a) Methanol

b) Ethanol

c) Ethylene glycol

d) Propylene glycol

18. What is the main product of photosynthesis?

a) Glucose

b) Oxygen

c) Carbon dioxide

d) Water

19. Which part of the brain controls voluntary actions?

a) Cerebellum

b) Medulla

c) Cerebrum

d) Hypothalamus

20. Which method is used by plants to reproduce without seeds?

a) Sexual reproduction

b) Asexual reproduction

c) Pollination

d) Fertilization

21. Which trait is dominant according to Mendel's laws?

a) Wrinkled seed

b) Tall plant

c) Green seed

d) Dwarf plant

22. What type of image is formed by a plane mirror?

a) Real and inverted

b) Real and upright

c) Virtual and inverted

d) Virtual and upright

23. Which part of the eye is responsible for controlling the amount of light entering the eye?

a) Retina

b) Iris

c) Cornea

d) Lens

24. What is the unit of electric resistance?

a) Volt

b) Ampere

c) Ohm

d) Watt

25. Which rule determines the direction of force on a currentcarrying conductor in a magnetic field?

a) Fleming's LeftHand Rule

b) Fleming's RightHand Rule

c) Ampere's Rule

d) Maxwell's Rule

26. Which is the most common greenhouse gas?

a) Oxygen

b) Methane

c) Carbon dioxide

d) Nitrogen

27. What type of reaction occurs when an acid reacts with a base?

a) Decomposition

b) Combustion

c) Neutralization

d) Redox

28. Which of the following acids is found in vinegar?

a) Acetic acid

b) Citric acid

c) Sulfuric acid

d) Hydrochloric acid

29. What is the main ore of aluminum?

a) Hematite

b) Bauxite

c) Galena

d) Magnetite

30. Which compound is a saturated hydrocarbon?

a) Ethene

b) Ethyne

c) Ethane

d) Benzene

31. Which part of the human digestive system is responsible for the absorption of nutrients?

a) Stomach

b) Small intestine

c) Large intestine

d) Esophagus

32. Which gland in the human body is referred to as the 'master gland'?

a) Thyroid

b) Adrenal

c) Pituitary

d) Pancreas

33. Which of the following is a mode of asexual reproduction in plants?

a) Fertilization

b) Pollination

c) Budding

d) Germination

34. Which scientist formulated the laws of inheritance?

a) Isaac Newton

b) Albert Einstein

c) Gregor Mendel

d) Charles Darwin

35. Which mirror is used in car headlights?

a) Plane mirror

b) Convex mirror

c) Concave mirror

d) Cylindrical mirror

36. Which phenomenon explains the splitting of white light into its constituent colors?

a) Refraction

b) Dispersion

c) Diffraction

d) Reflection

37. Which law states that the current through a conductor between two points is directly proportional to the voltage across the two points?

a) Newton's Law

b) Coulomb's Law

c) Ohm's Law

d) Faraday's Law

38. Which device is used to measure electric current?

a) Voltmeter

b) Ammeter

c) Galvanometer

d) Ohmmeter

39. Which of the following is a renewable source of energy?

a) Coal

b) Natural gas

c) Wind energy

d) Nuclear energy

40. Which of the following is a chemical change?

a) Melting of ice

b) Dissolving sugar in water

c) Rusting of iron

d) Boiling of water

41. What is formed when an acid reacts with a metal?

a) Salt and water

b) Salt and hydrogen gas

c) Salt and carbon dioxide

d) Salt and oxygen gas

42. Which metal is liquid at room temperature?

a) Mercury

b) Gold

c) Silver

d) Iron

43. What is the general formula for alkanes?

a) C_nH_{2n+2}

b) C_nH_{2n}

c) C_nH_{2n-2}

d) C_nH_{n+2}

44. Which organelle is known as the 'powerhouse of the cell'?

a) Nucleus

b) Ribosome

c) Mitochondrion

d) Endoplasmic reticulum

45. Which hormone is responsible for the fightorflight response?

a) Insulin

b) Adrenaline

c) Melatonin

d) Estrogen

46. Which process results in the formation of a zygote?

a) Pollination

b) Fertilization

c) Germination

d) Spore formation

47. Which concept did Mendel use to explain how traits are inherited?

a) Chromosomal theory

b) Germplasm theory

c) Particulate theory

d) Pangenesis

48. What is the angle of incidence equal to in the law of reflection?

a) Angle of reflection

b) Angle of refraction

c) Critical angle

d) Angle of emergence

49. Which lens is used to correct myopia (nearsightedness)?

a) Convex lens

b) Concave lens

c) Cylindrical lens

d) Bifocal lens

50. What is the formula for calculating electric power?

a) $P = IV$

b) $P = I^2R$

c) $P = V^2/R$

d) All of the above

Questions: Part 2

51. Which device is used to convert electrical energy into mechanical energy?

a) Transformer

b) Generator

c) Motor

d) Battery

52. Which layer of the atmosphere contains the ozone layer?

a) Troposphere

b) Stratosphere

c) Mesosphere

d) Thermosphere

53. Which type of reaction involves the exchange of ions between two compounds?

a) Synthesis reaction

b) Decomposition reaction

c) Double displacement reaction

d) Single displacement reaction

54. Which base is used in the manufacture of soap?

a) Sodium hydroxide

b) Ammonium hydroxide

c) Calcium hydroxide

d) Potassium hydroxide

55. Which metal is used in the galvanization process?

a) Copper

b) Zinc

c) Iron

d) Silver

56. Which functional group is present in alcohols?

a) CHO

b) OH

c) COOH

d) NH_2

57. Which part of the cell is responsible for protein synthesis?

a) Nucleus

b) Ribosomes

c) Mitochondria

d) Golgi apparatus

58. Which hormone regulates the sleepwake cycle?

a) Insulin

b) Adrenaline

c) Thyroxine

d) Melatonin

59. Which process is essential for the continuation of species?

a) Nutrition

b) Reproduction

c) Respiration

d) Excretion

60. Which law of inheritance states that alleles segregate independently of each other?

a) Law of Dominance

b) Law of Segregation

c) Law of Independent Assortment

d) Law of Recombination

61. Which mirror is used as a rearview mirror in vehicles?

a) Plane mirror

b) Convex mirror

c) Concave mirror

d) Spherical mirror

62. What causes the formation of a rainbow?

a) Reflection

b) Refraction and dispersion

c) Diffraction

d) Polarization

63. Which factor affects the resistance of a conductor?

a) Length

b) Crosssectional area

c) Temperature

d) All of the above

64. Which phenomenon causes the needle of a compass to point north?

a) Earth's gravitational field

b) Earth's magnetic field

c) Earth's electric field

d) None of the above

65. Which human activity contributes most to deforestation?

a) Urbanization

b) Industrialization

c) Agriculture

d) Mining

66. Which of the following is a characteristic of a redox reaction?

a) Transfer of protons

b) Transfer of electrons

c) Formation of precipitate

d) Release of gas

67. Which indicator turns red in an acidic solution?

a) Methyl orange

b) Phenolphthalein

c) Litmus

d) Bromothymol blue

68. What is the process of coating iron with zinc to prevent rusting called?

a) Electroplating

b) Anodizing

c) Galvanization

d) Corrosion

69. Which of the following is a hydrocarbon with a triple bond?

a) Ethane

b) Ethene

c) Ethyne

d) Benzene

70. Which blood vessel carries oxygenated blood from the lungs to the heart?

a) Pulmonary artery

b) Pulmonary vein

c) Aorta

d) Vena cava

71. Which hormone is involved in the regulation of calcium levels in the blood?

a) Insulin

b) Adrenaline

c) Parathyroid hormone

d) Thyroxine

72. Which reproductive process involves the fusion of male and female gametes?

a) Asexual reproduction

b) Binary fission

c) Fertilization

d) Budding

73. Which concept did Mendel use to describe how traits are passed from parents to offspring?

a) Blending inheritance

b) Gene theory

c) Chromosome theory

d) Independent assortment

74. Which property of light is responsible for the formation of shadows?

a) Reflection

b) Refraction

c) Rectilinear propagation

d) Diffraction

75. Which part of the eye adjusts its shape to focus light on the retina?

a) Cornea

b) Iris

c) Lens

d) Pupil

76. Which factor does not affect the resistance of a conductor?

a) Material of the conductor

b) Temperature of the conductor

c) Shape of the conductor

d) Length of the conductor

77. Which device is used to increase or decrease the voltage of an alternating current?

a) Motor

b) Transformer

c) Generator

d) Rectifier

78. Which of the following is an example of a biodegradable substance?

a) Plastic

b) Glass

c) Paper

d) Aluminum

79. Which reaction type is represented by the equation: $AB \rightarrow A + B$?

a) Synthesis reaction

b) Decomposition reaction

c) Double displacement reaction

d) Combustion reaction

80. Which acid is produced in the stomach to aid digestion?

a) Sulfuric acid

b) Hydrochloric acid

c) Nitric acid

d) Acetic acid

81. Which metal is extracted from the ore bauxite?

a) Iron

b) Copper

c) Aluminum

d) Zinc

82. What is the molecular formula of methane?

a) CH_4

b) C_2H_4

c) C_2H_2

d) C_3H_8

83. Which organ is the primary site for nutrient absorption in humans?

a) Stomach

b) Small intestine

c) Large intestine

d) Liver

84. Which hormone regulates the menstrual cycle in females?

a) Insulin

b) Estrogen

c) Testosterone

d) Thyroxine

85. Which process involves the transfer of pollen from the anther to the stigma of a flower?

a) Germination

b) Pollination

c) Fertilization

d) Vegetative propagation

86. Which of Mendel's laws explains why offspring may have combinations of traits not seen in their parents?

a) Law of Dominance

b) Law of Segregation

c) Law of Independent Assortment

d) Law of Recombination

87. Which phenomenon occurs when light passes from one medium to another and bends?

a) Reflection

b) Refraction

c) Dispersion

d) Diffraction

88. Which part of the eye contains photoreceptor cells?

a) Cornea

b) Retina

c) Lens

d) Sclera

89. Which physical quantity is directly proportional to the square of the current?

a) Voltage

b) Resistance

c) Power

d) Conductance

90. Which device converts mechanical energy into electrical energy?

a) Motor

b) Generator

c) Transformer

d) Rectifier

91. Which of the following is a nonrenewable resource?

a) Solar energy

b) Wind energy

c) Coal

d) Hydropower

92. Which term describes a reaction where heat is absorbed from the surroundings?

a) Exothermic

b) Endothermic

c) Combustion

d) Neutralization

93. Which base is commonly found in household cleaning products?

a) Sodium hydroxide

b) Potassium hydroxide

c) Ammonium hydroxide

d) Calcium hydroxide

94. Which metal reacts vigorously with water to form an alkaline solution?

a) Magnesium

b) Copper

c) Sodium

d) Zinc

95. Which hydrocarbon is a major component of natural gas?

a) Ethane

b) Propane

c) Butane

d) Methane

96. Which part of the respiratory system is the site of gas exchange?

a) Trachea

b) Bronchi

c) Alveoli

d) Diaphragm

97. Which hormone is secreted by the pancreas?

a) Adrenaline

b) Insulin

c) Thyroxine

d) Glucagon

98. Which method of asexual reproduction is common in yeast?

a) Binary fission

b) Budding

c) Fragmentation

d) Spore formation

99. Which genetic concept explains the variation in traits observed in a population?

a) Gene linkage

b) Genetic drift

c) Mutation

d) Natural selection

100. Which type of image is formed by a convex lens when an object is placed beyond twice the focal length?

a) Real, inverted, and smaller

b) Real, inverted, and larger

c) Virtual, upright, and larger

d) Virtual, upright, and smaller

Questions: Part 3

101. Which color of light has the shortest wavelength?

a) Red

b) Green

c) Blue

d) Violet

102. Which material is used as the filament in an incandescent light bulb?

a) Copper

b) Aluminum

c) Tungsten

d) Iron

103. Which rule is applied to determine the direction of magnetic field lines around a currentcarrying conductor?

a) Fleming's LeftHand Rule

b) RightHand Thumb Rule

c) Ampere's Rule

d) Lenz's Law

104. Which of the following contributes most to global warming?

a) Water vapor

b) Carbon dioxide

c) Methane

d) Nitrous oxide

105. Which chemical reaction represents a combination reaction?

a) $2H_2 + O_2 \rightarrow 2H_2O$

b) $H_2O_2 \rightarrow H_2O + O_2$

c) $NaCl + AgNO_3 \rightarrow NaNO_3 + AgCl$

d) $HCl + NaOH \rightarrow NaCl + H_2O$

106. Which of the following acids is a weak acid?

a) Hydrochloric acid

b) Sulfuric acid

c) Acetic acid

d) Nitric acid

107. Which metal is used in the construction of aircraft due to its lightweight and strength?

a) Iron

b) Copper

c) Aluminum

d) Zinc

108. Which of the following hydrocarbons is an unsaturated compound?

a) Propane

b) Butane

c) Ethene

d) Methane

109. Which process in the human body breaks down glucose to release energy?

a) Photosynthesis

b) Respiration

c) Fermentation

d) Digestion

110. Which endocrine gland is located at the base of the brain?

a) Thyroid gland

b) Adrenal gland

c) Pituitary gland

d) Pancreas

111. Which method of reproduction involves the production of seeds in plants?

a) Asexual reproduction

b) Vegetative propagation

c) Sexual reproduction

d) Spore formation

112. Which genetic cross represents a monohybrid cross?

a) TT x tt

b) Tt x Tt

c) TT x TT

d) Tt x tt

113. Which optical instrument uses a convex lens to magnify small objects?

a) Microscope

b) Telescope

c) Periscope

d) Binoculars

114. Which phenomenon explains why the sky appears blue?

a) Reflection

b) Refraction

c) Scattering

d) Dispersion

115. Which electrical component is designed to resist the flow of current?

a) Resistor

b) Capacitor

c) Inductor

d) Transformer

116. Which device uses electromagnetic induction to generate electricity?

a) Battery

b) Motor

c) Generator

d) Transformer

117. Which substance is a major contributor to acid rain?

a) Carbon monoxide

b) Sulfur dioxide

c) Methane

d) Oxygen

118. Which reaction occurs when an acid reacts with a carbonate?

a) Formation of salt and water

b) Formation of salt, water, and carbon dioxide

c) Formation of hydrogen gas

d) No reaction occurs

119. Which of the following bases is used in baking to make cakes rise?

a) Sodium hydroxide

b) Potassium hydroxide

c) Ammonium hydroxide

d) Sodium bicarbonate

120. Which metal does not corrode easily in moist air?

a) Iron

b) Copper

c) Silver

d) Gold

121. Which element is found in the largest amount in the Earth's crust?

a) Iron

b) Silicon

c) Aluminum

d) Oxygen

122. Which compound is a member of the alcohol family?

a) Methanol

b) Methanal

c) Methanoic acid

d) Methane

123. Which organ in the human body is primarily responsible for filtering waste from the blood?

a) Liver

b) Heart

c) Kidney

d) Pancreas

124. Which hormone regulates metabolism in the human body?

a) Insulin

b) Glucagon

c) Thyroxine

d) Estrogen

125. Which process in plants is responsible for producing food from sunlight?

a) Respiration

b) Germination

c) Photosynthesis

d) Transpiration

126. Which genetic disorder is caused by an extra copy of chromosome 21?

a) Down syndrome

b) Turner syndrome

c) Klinefelter syndrome

d) Cystic fibrosis

127. Which type of lens is used to correct hypermetropia (farsightedness)?

a) Convex lens

b) Concave lens

c) Cylindrical lens

d) Planoconcave lens

128. Which factor affects the focal length of a lens?

a) Thickness of the lens

b) Refractive index of the lens material

c) Radius of curvature of the lens surfaces

d) All of the above

129. Which of the following materials is a good conductor of electricity?

a) Glass

b) Rubber

c) Copper

d) Wood

130. Which rule is used to determine the direction of induced current in a conductor?

a) Fleming's RightHand Rule

b) Fleming's LeftHand Rule

c) Ampere's Law

d) Lenz's Law

131. Which human activity is the largest source of carbon dioxide emissions?

a) Deforestation

b) Transportation

c) Agriculture

d) Industrial processes

132. Which of the following is a chemical property?

a) Melting point

b) Boiling point

c) Flammability

d) Density

133. Which type of reaction occurs when two solutions are mixed and a solid forms?

a) Precipitation reaction

b) Redox reaction

c) Neutralization reaction

d) Decomposition reaction

134. Which acid is a major component of acid rain?

a) Hydrochloric acid

b) Acetic acid

c) Sulfuric acid

d) Nitric acid

135. Which metal is most commonly used in electrical wiring?

a) Gold

b) Iron

c) Copper

d) Aluminum

136. Which organic compound is the simplest alcohol?

a) Methanol

b) Ethanol

c) Propanol

d) Butanol

137. Which part of the circulatory system carries blood away from the heart?

a) Veins

b) Arteries

c) Capillaries

d) Lymph vessels

138. Which hormone is known as the 'stress hormone'?

a) Insulin

b) Cortisol

c) Adrenaline

d) Thyroxine

139. Which process involves the movement of water molecules through a semipermeable membrane?

a) Diffusion

b) Osmosis

c) Active transport

d) Endocytosis

140. Which genetic principle explains the variety of phenotypes observed in a population?

a) Mutation

b) Genetic drift

c) Natural selection

d) Recombination

141. Which type of mirror is commonly used in shaving mirrors?

a) Plane mirror

b) Concave mirror

c) Convex mirror

d) Spherical mirror

142. Which property of light is responsible for the formation of mirages?

a) Reflection

b) Refraction

c) Diffraction

d) Dispersion

143. Which unit is used to measure the electric potential difference?

a) Volt

b) Ampere

c) Ohm

d) Joule

144. Which device is used to measure the resistance of a material?

a) Voltmeter

b) Ammeter

c) Ohmmeter

d) Galvanometer

145. Which gas is primarily responsible for the greenhouse effect?

a) Nitrogen

b) Oxygen

c) Carbon dioxide

d) Argon

146. Which type of chemical bond involves the sharing of electron pairs between atoms?

a) Ionic bond

b) Covalent bond

c) Metallic bond

d) Hydrogen bond

147. Which base is used in antacids to neutralize stomach acid?

a) Sodium hydroxide

b) Calcium carbonate

c) Magnesium hydroxide

d) Potassium hydroxide

148. Which metal is extracted from the mineral hematite?

a) Copper

b) Iron

c) Aluminum

d) Zinc

149. Which type of organic compound contains a carbonyl group attached to a hydrogen atom?

a) Alcohol

b) Aldehyde

c) Ketone

d) Carboxylic acid

150. Which part of the brain is responsible for coordinating voluntary movements?

a) Cerebrum

b) Cerebellum

c) Medulla oblongata

d) Hypothalamus

Answers

1. b) $2H_2 + O_2 \rightarrow 2H_2O$

2. b) 7

3. c) Sodium

4. b) Graphite

5. b) Respiration

6. a) Insulin

7. a) Binary fission

8. b) Gregor Mendel

9. c) Law of Reflection

10. b) Seeing

11. b) Ampere

12. b) Magnetic force

13. b) Decomposition

14. a) Mass

15. c) Sour taste

16. a) CaO

17. c) Ethylene glycol

18. a) Glucose

19. c) Cerebrum

20. b) Asexual reproduction

21. b) Tall plant

22. d) Virtual and upright

23. b) Iris

24. c) Ohm

25. a) Fleming's LeftHand Rule

26. c) Carbon dioxide

27. c) Neutralization

28. a) Acetic acid

29. b) Bauxite

30. c) Ethane

31. b) Small intestine

32. c) Pituitary

33. c) Budding

34. c) Gregor Mendel

35. c) Concave mirror

36. b) Dispersion

37. c) Ohm's Law

38. b) Ammeter

39. c) Wind energy

40. c) Rusting of iron

41. b) Salt and hydrogen gas

42. a) Mercury

43. a) C_nH_{2n+2}

44. c) Mitochondrion

45. b) Adrenaline

46. b) Fertilization

47. c) Particulate theory

48. a) Angle of reflection

49. b) Concave lens

50. d) All of the above

51. c) Motor

52. b) Stratosphere

53. c) Double displacement reaction

54. a) Sodium hydroxide

55. b) Zinc

56. b) OH

57. b) Ribosomes

58. d) Melatonin

59. b) Reproduction

60. c) Law of Independent Assortment

61. b) Convex mirror

62. b) Refraction and dispersion

63. d) All of the above

64. b) Earth's magnetic field

65. c) Agriculture

66. b) Transfer of electrons

67. c) Litmus

68. c) Galvanization

69. c) Ethyne

70. b) Pulmonary vein

71. c) Parathyroid hormone

72. c) Fertilization

73. b) Gene theory

74. c) Rectilinear propagation

75. c) Lens

76. c) Shape of the conductor

77. b) Transformer

78. c) Paper

79. b) Decomposition reaction

80. b) Hydrochloric acid

81. c) Aluminum

82. a) CH_4

83. b) Small intestine

84. b) Estrogen

85. b) Pollination

86. c) Law of Independent Assortment

87. b) Refraction

88. b) Retina

89. c) Power

90. b) Generator

91. c) Coal

92. b) Endothermic

93. c) Ammonium hydroxide

94. c) Sodium

95. d) Methane

96. c) Alveoli

97. b) Insulin

98. b) Budding

99. c) Mutation

100. b) Real, inverted, and larger

101. d) Violet

102. c) Tungsten

103. b) RightHand Thumb Rule

104. b) Carbon dioxide

105. a) $2H_2 + O_2 \rightarrow 2H_2O$

106. c) Acetic acid

107. c) Aluminum

108. c) Ethene

109. b) Respiration

110. c) Pituitary gland

111. c) Sexual reproduction

112. a) TT x tt

113. a) Microscope

114. c) Scattering

115. a) Resistor

116. c) Generator

117. b) Sulfur dioxide

118. b) Formation of salt, water, and carbon dioxide

119. d) Sodium bicarbonate

120. d) Gold

121. d) Oxygen

122. a) Methanol

123. c) Kidney

124. c) Thyroxine

125. c) Photosynthesis

126. a) Down syndrome

127. a) Convex lens

128. d) All of the above

129. c) Copper

130. a) Fleming's RightHand Rule

131. d) Industrial processes

132. c) Flammability

133. a) Precipitation reaction

134. c) Sulfuric acid

135. c) Copper

136. a) Methanol

137. b) Arteries

138. b) Cortisol

139. b) Osmosis

140. d) Recombination

141. b) Concave mirror

142. b) Refraction

143. a) Volt

144. c) Ohmmeter

145. c) Carbon dioxide

146. b) Covalent bond

147. c) Magnesium hydroxide

148. b) Iron

149. b) Aldehyde

150. b) Cerebellum

Printed at Repro India Ltd.